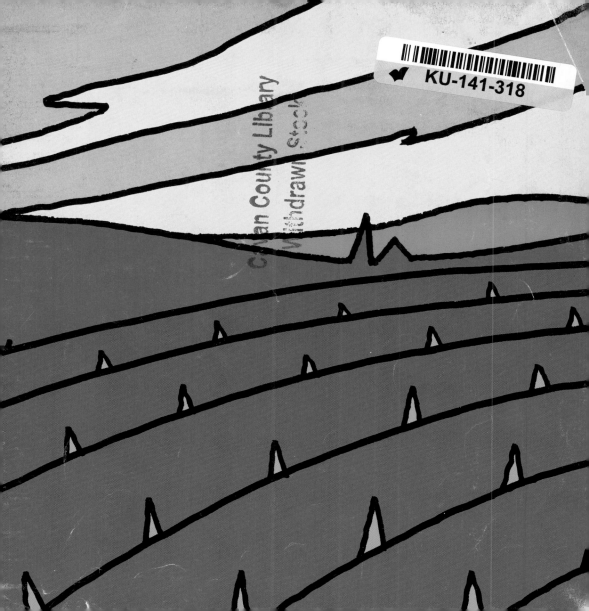

The Jan Pieńkowski Nursery Books:
ABC, Colours, Faces, Farm, Food, Homes,
Numbers, Shapes, Sizes, Time, Weather, Zoo

for Alexander

William Heinemann Ltd
an imprint of Reed Books Ltd
Michelin House, 81 Fulham Road, London SW3 6RB
and Auckland, Melbourne, Singapore and Toronto

First published 1985
Text and Illustrations copyright © Jan Pieńkowski 1985
Reprinted 1986, 1992, 1995
0 434 95651 1

Lettering by Caroline Austin
Produced by Mandarin Offset Ltd
Printed and bound in Hong Kong

FARM

Jan Pieńkowski

HEINEMANN : LONDON

farmers

lambs

bull

tractor

horse

cock

hen

chicks

pigs

sow

piglets

cat

ducks

geese

scarecrow

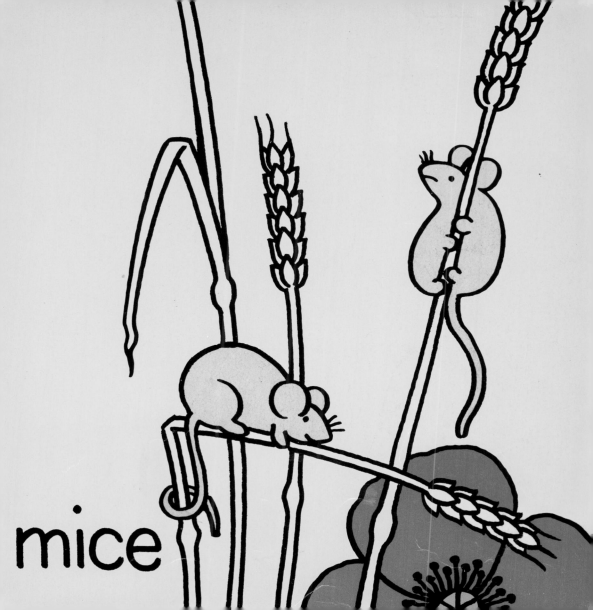

mice

rabbits

donkey

turkey

goat

what animal?